Canon EOS Rebel T7 guide

Step-by-Step Camera Settings, Modes, and DSLR Photography Basics for Beginners

Marc T. Billy

Contents

Introduction

There's something sacred about holding a real camera in your hands. Not a phone. Not an app. A machine with dials, glass, and just enough complexity to remind you: *this is real photography.*

But then you turn it on.

Suddenly you're staring at menus that look like code, buttons that don't respond the way you expect, and a hundred acronyms Canon never bothered to explain. ISO? AF? TV mode? Grid display? White balance? And just like that, the excitement that led you to buy the Rebel T7 starts to evaporate—replaced by the creeping fear that maybe, just maybe, this camera is too much for you.

Let me promise you something upfront:
It's not.

In fact, the Canon EOS Rebel T7 might be the best tool you'll ever own for learning photography the *right* way. It was built for everyday people—parents, travelers, creators, students, hobbyists, and absolute beginners—who want more control, better photos, and the confidence to shoot anything. But here's the truth: the manual that came with your camera? It wasn't written with you in mind.

This book is.

We're not going to walk through boring spec sheets or talk to you like a camera engineer. We're going to walk beside you—step by step—as if you just picked this camera up for the first time and said, *"Okay, now what?"*

You'll learn what every button does without getting overwhelmed. You'll finally understand what shutter speed, aperture, and ISO *mean*—not just technically, but practically. You'll discover how to shoot in manual mode without fear, how to fix blurry or dark photos instantly, how to set up your camera like a pro in 5 minutes, and how to actually take the kinds of pictures you imagined when you bought this thing.

You'll also unlock features and tricks Canon didn't tell you about. Real-world shortcuts. Autofocus tweaks. A lens setting that makes colors pop. Menu hacks that can literally rescue a bad photo before you even take it. And don't worry—you don't need to know a thing about photography right now. That's what I'm here for.

This book was made to feel like a conversation—not a lecture. No filler. No robotic jargon. Just clean, clear steps and human language you'll *actually* understand. It's like having a patient friend who speaks your language and knows this camera inside out, walking you through each setting, each dial, each moment—until something clicks. And it *will* click.

Because photography isn't about having perfect technique.
It's about presence.

It's about standing in front of something that moves you—a child's laugh, a city at dusk, a stranger in beautiful light—and knowing *how* to respond.
How to freeze that feeling forever.

This isn't just a camera manual. It's your *unlock.*
The missing bridge between frustration and confidence. Between autopilot and intentional creation. Between "I hope this turns out okay" and "I *know* exactly what I'm doing."

So whether this is your first camera or just your first time really learning it...
Welcome.
Let's begin.

Chapter 1

Meet Your Canon EOS Rebel T7: A Camera Built for Learning

If you're holding a Canon EOS Rebel T7 right now, you're holding one of the most forgiving, reliable, and beginner-friendly DSLR cameras ever made. That's not hype—it's truth. While it may not have the flashiest specs or the fanciest price tag, the T7 has quietly become the *first camera* for millions of people around the world. And that matters.

Because the first camera you own—the one you actually *learn* on—can either ignite your passion for photography or bury it under menus and frustration. Canon got something right with the Rebel T7. Now it's your turn to unlock that potential.

But let's be honest. Right now, that camera might still feel more like a puzzle than a tool.

There's a screen full of symbols you don't understand. The buttons seem to do five things at once. You're not even sure if your photos are turning out okay. And when you opened the official Canon manual, you probably closed it just as fast.

This chapter is your reset button.

We're going to break down:

- **What the Rebel T7 actually is**—and why Canon designed it the way they did
- **Why it's ideal for first-time photographers**, even if it doesn't look like it yet
- What's *missing* from the default experience that this book will give you
- And how to think like a photographer—without any prior experience

Let's meet your camera in a way Canon never introduced it.

The Big Picture: What the Canon EOS Rebel T7 Actually *Is*

The Rebel T7 (also known globally as the **EOS 2000D**) is an **entry-level DSLR camera** designed to be a learning bridge between point-and-shoot or smartphone photography and professional image-making. It's the gateway drug to real photography.

What makes it a DSLR?

- **DSLR** stands for *Digital Single Lens Reflex*
- It uses a mirror and prism system to show you the real scene through the viewfinder—not a digital preview like mirrorless cameras
- That mirror mechanism gives you fast autofocus, natural eye-level composition, and a sense of classic photo-taking

that smartphone cameras can't replicate

The T7 is built around

- **A 24.1 megapixel APS-C CMOS sensor** – large enough to give you excellent image quality in all lighting conditions
- **Canon's DIGIC 4+ image processor** – an engine that balances speed, image sharpness, and color rendering
- **An EF/EF-S lens mount** – compatible with dozens of Canon lenses (and third-party options) to grow with you

But specs are boring if you don't know what they *mean.*
Here's the real takeaway: **the T7 is built to teach you how to see.** It forces you to slow down just enough to understand *how* photography works—but not so much that you get lost in complexity.

Who the Rebel T7 Was *Really* Designed For

Canon never marketed this camera to pros. They didn't load it with overwhelming menus or complex configurations. Instead, they made the T7 for

- **Beginners:** People holding a DSLR for the very first time
- **Seniors & retirees:** Those who want to learn photography as a passion project—without the chaos of modern tech
- **Parents:** Trying to document childhood, holidays, or family memories with better quality

- **Vloggers & creators:** On a budget but seeking better visuals than a smartphone can deliver
- **Travelers:** Wanting a lightweight, quality DSLR that won't feel like a burden

If that sounds like you, good. This book was written with you in mind.

What the Canon Manual *Doesn't* Tell You (That You Desperately Need)

Here's the problem: most people give up on the Rebel T7 before they ever unlock its full power. Not because it's a bad camera—it isn't—but because Canon didn't give you a human guide.

The official manual is

- Over 140 pages long
- Filled with unexplained icons, settings, and jargon
- Not structured in a way that matches how people actually *learn*

And if you're like most new users, you've probably:

- Left the camera in Auto mode, hoping for the best
- Taken a few decent shots, but couldn't repeat them
- Struggled with blurry photos, dark images, or slow focus
- Wondered if the problem is *you* (spoiler: it's not)

Here's the truth: photography isn't about knowing every feature—it's about knowing which features matter to you. This book will teach you that. And it starts here.

Think Like a Photographer (Even If You've Never Taken a "Real" Photo)

Let me shift your mindset right now.

Most beginners think photography is about pressing a button at the right time.

But real photography is about **seeing**—not just looking. It's about

- Recognizing light and how it shapes your subject
- Understanding what your viewer will notice first
- Making deliberate choices that guide the eye
- Capturing emotion, not just information

And here's the good news: **the Rebel T7 forces you to slow down and notice these things.**
It's not about blasting off 200 photos in 3 seconds. It's about being intentional. About telling a story with light, angle, timing, and mood. About making the viewer *feel something*.

> You don't need perfect technique.
> You need presence.

The rest—buttons, modes, menus, tricks—that's what this book is for.

What Happens in This Book (And What You'll Gain)

Each chapter is structured to build on the last. You'll go from:

- Wondering what a dial does…
  → To switching modes with confidence

- Guessing which setting to change…
  → To tweaking your camera like a pro

- Hoping for a decent shot…
  → To *knowing* you got the shot

This is more than a guide.
It's your map. Your secret weapon. Your creative permission slip.

You're not just learning how to use a camera.
You're learning how to *see your world differently*—and save it forever, one frame at a time.

Are you ready?

Because in the next chapter, we're going to unbox that camera again.
Not physically—but *properly*.
And we're going to set it up so that *you*, not the camera, are in control.

Let's keep going.

Chapter 2

Unboxing & Setup: Camera Ready in 10 Minutes

Let's reset. Whether your Canon EOS Rebel T7 is freshly unboxed or has been sitting in a drawer for a few weeks (or months), we're going to walk through the setup together like it's day one. This chapter is about more than just plugging in batteries and turning the camera on—it's about understanding *why* each piece matters, what most people get wrong, and how to configure your camera so it works *with* you, not against you.

You don't need to be technical. You don't need to know anything yet. You just need to be here, with your camera in front of you. That's more than enough.

What's in the Box (And What's Missing)

When you open your Canon Rebel T7 box, here's what you should find:

- Canon EOS Rebel T7 Camera Body
- Canon EF-S 18–55mm f/3.5–5.6 IS II Kit Lens
- LP-E10 Rechargeable Battery Pack

- LC-E10 Battery Charger
- Camera Strap
- Interface Cable (usually micro USB)
- Instruction Manual (physical or digital access)

But what they *don't* tell you is just as important:

What's Not Included (but absolutely essential):

- **SD Memory Card** — Your camera won't take photos or record video without one
- **Protective Camera Bag**—For travel or storage
- **Microfiber Cloth** — To safely clean your lens
- **Screen Protector** — The LCD can scratch more easily than you'd think
- **Lens Cap Keeper** — Because you *will* lose it otherwise
- **Spare Battery** — Especially for trips or long shoots
- **Basic Tripod** — Essential for low-light, night, or self-portrait shots

These aren't luxury items. They're practical tools that transform frustration into smooth, enjoyable shooting. If you haven't picked them up yet, consider this your short-term shopping list.

Step 1: Insert Battery and Memory Card

1. **Flip the camera over.** Open the battery compartment door.

2. **Slide in the LP-E10 battery.** It should click firmly into place.

3. **Insert your SD memory card.** Make sure it clicks into the slot above the battery (label side toward the back of the camera).

4. Close the battery door and switch the power **ON.**

Tip: Use a **Class 10 UHS-I SD card** (at least 32GB) for smooth photo and video performance. Avoid using ultra-cheap cards from gas stations or random online stores—they fail often.

Step 2: Attach the Lens (Properly)

1. **Remove the body cap** from your camera and the rear cap from the lens.

2. **Line up the white square (or red dot)** on the lens with the matching mark on the camera.

    - *White square = EF-S lens (standard for T7)*
    - *Red dot = EF lens (also compatible)*

3. Rotate the lens clockwise until it clicks securely into place.

Don't force it. If it doesn't turn easily, it's misaligned. Twist gently and confidently.

Step 3: Set the Language, Time & Date

The first time you turn your camera on, it'll ask for some basics:

- Select **Language** (English, unless you're multilingual)
- Set **Time Zone** (your location)
- Input the **Date & Time** (accurate timestamps help organize photos)
- Choose **Date Format** (YYYY/MM/DD is safest for file sorting)

Tip: Setting this correctly will help you avoid misdated photo folders—especially helpful when importing photos to your computer or cloud.

Step 4: Customize Basic Camera Settings

Canon's factory defaults are *okay*, but let's make a few essential adjustments right away.

1. **Press the Menu button** (top-left on the back).
2. Scroll using the arrow keys or main dial.

Now, tweak these:

- **Auto Power Off** → Set to 1 min (saves battery)
- **Image Review** → 2 sec (lets you quickly check your photo)

- **Beep** → OFF (so your camera doesn't chirp constantly)
- **Date/Time** → Double-check accuracy
- **Image Quality** → Set to **"L" JPEG** for now (we'll get into RAW later)

Optional but Recommended:

- **Grid Display** → Enable (helps composition)
- **Highlight Alert** → Enable (warns you if parts of your image are overexposed)
- **Lens Aberration Correction** → Enable both (cleaner images without editing)

Step 5: Set Up Wi-Fi & Bluetooth (Optional But Worth It)

The Canon Rebel T7 includes built-in **Wi-Fi** for image transfer. It does *not* have Bluetooth or a touchscreen.

To connect to your phone:

1. Download the **Canon Camera Connect** app (iOS/Android).
2. On your camera, go to
Menu > Wi-Fi Function > Connect to Smartphone
3. Follow the pairing steps shown on-screen.
4. Complete pairing via the app.

Why this matters:
You'll be able to wirelessly:

- Transfer images to your phone
- Use your phone as a remote shutter (great for self-portraits!)
- Instantly share your shots to social media or cloud

Step 6: Format Your SD Card

Even if your memory card is brand new, format it using your camera to make sure it's properly optimized.

- Menu > Format Card > OK

This clears any hidden files and prepares the card for Canon's file structure.

> **Warning:** Formatting erases everything. Never format a card that holds photos you haven't backed up.

Step 7: Practice Holding the Camera Correctly

It sounds basic—but how you hold your Rebel T7 directly affects sharpness and focus.

- **Right Hand:** Grips the camera body firmly, index finger over the shutter button.
- **Left Hand:** Cradles the lens from below, not the side.
- **Arms In:** Tuck your elbows toward your body to minimize shake.

- **Feet Shoulder-Width Apart:** You are the tripod until you get a real one.

This is your first step to avoiding blurry images—no fancy settings required.

What You've Just Done (That Most Beginners Don't)

By completing this setup:

- You've optimized your T7 for real-world shooting—not factory defaults
- You've activated hidden settings that make photography easier
- You've avoided the most common beginner mistakes (like skipping card formatting or never learning lens attachment)
- You've created a clean foundation for learning *on your terms*

You've also officially taken control of your camera.
From here on out, you're no longer a passive shooter—you're an active image maker.

This wasn't just a setup. It was a small transformation.
The kind that sets you apart from 90% of Rebel T7 owners who never touch a single menu after turning the camera on.

Chapter 3

BUTTONS, DIALS & WHAT THEY ACTUALLY DO (AND HOW TO USE THEM WITHOUT GUESSING)

The Canon EOS Rebel T7 may look like it's covered in mysterious symbols and buttons—but once you understand what each one does, the fear vanishes. This chapter is about **eliminating confusion.** You'll no longer press buttons wondering what might happen.

Instead, by the end of this chapter, you'll know:

- What each button and dial *really* controls
- Which ones matter right now, and which you can ignore until later
- How to combine buttons to access hidden or advanced functions
- What to press when something goes wrong (blur, exposure, weird colors)
- And why button placement actually helps you shoot faster and smarter

This isn't memorization. This is orientation.
And once you're oriented, you become confident.

Let's Mentally Map Your Camera First

Imagine holding your Canon T7 in your hands. Let's walk around it piece by piece like we're in a cockpit for the first time—not to be overwhelmed, but to *learn the controls.*

We'll go from **top** → **back** → **front** → **side** in a simple, clear way.

Top of the Camera: Where Power Meets Precision

Here's where the **most essential shooting tools** are located.

1. Mode Dial

This is the big circular dial with letters and icons.

It's where you choose *how* the camera behaves—from Auto to Manual.

- **Scene Intelligent Auto (green box):** Point-and-shoot simplicity
- **P (Program Auto):** Like Auto, but lets you tweak certain things
- **Tv (Shutter Priority):** You set shutter speed, camera picks the rest
- **Av (Aperture Priority):** You choose depth of field, camera adjusts the rest

- **M (Manual Mode):** You control everything—ISO, shutter, aperture
- **Scene Modes (Portrait, Landscape, etc.):** Auto presets optimized for specific situations

You'll learn when to use each mode in later chapters—but know this: **Manual Mode isn't as scary as it looks.**

2. Main Dial (Near the Shutter Button)

The silver scroll wheel you turn with your index finger.

- Changes settings depending on what mode you're in.
- For example:
  - In **Manual Mode**, it adjusts *shutter speed*
  - In **Av Mode**, it adjusts *aperture*
  - In menus, it scrolls through options

Tip: Think of this as your "fine-tune" control.

3. Shutter Button

The button you press to take a photo.

- Press halfway = Focus
- Press fully = Capture photo

This two-step action is powerful. Holding halfway helps you **lock focus** before shooting.

4. ISO Button

(Just behind the main dial, marked as "ISO")

- Press it to change your ISO setting—how sensitive your camera is to light.
- Lower ISO = cleaner image in bright light
- Higher ISO is better for low light, but adds grain/noise

5. Flash Button

Pops up the flash when in certain modes.

- Only available in modes that allow flash (Auto, Portrait, etc.)
- Doesn't work in Manual Mode unless specifically enabled

Back of the Camera: Your Command Center

This is where you review, adjust, and control what's happening on-screen.

6. Menu Button

(Top-left corner)

- Opens the main camera menu (where all the deeper settings live)

We'll walk through this in detail later—but think of this as your camera's "system settings."

7. Info Button

Cycles through what appears on your screen.

- Useful for cleaning up or customizing your LCD view
- Hit this when your screen feels cluttered—or when it's too empty

8. Playback Button (Blue Arrow)

Lets you view your photos on the screen.

- Use the arrows or dial to scroll
- Press the **trash icon** to delete (bottom-right corner)

9. Directional Pad / Arrow Keys

Helps navigate menus and move focus points.

- Press up = Change ISO (shortcut)
- Press right = Change AF mode
- Press down = Change white balance
- Press left = Drive/self-timer

Each direction leads to a *shortcut* setting. These buttons are hidden gems.

10. Q Button (Quick Control)

The most underrated button on your camera.

- Press this while shooting to access a **shortcut menu** on-screen.
- Here, you can quickly adjust:
  - ISO
  - White Balance
  - Autofocus Mode
  - Image Quality
  - Exposure Compensation

Use this button **often.** It keeps you out of the deep menus and speeds up your flow.

11. Set Button (Center of Arrow Pad)

Select any option you've highlighted.

- Think of it as your "OK" or "confirm" button.

Front of the Camera: Sensor Side Support

You won't use these often, but they matter.

12. Lens Release Button

Located next to the lens mount (on the right side when facing camera)

- Press this to **detach your lens** safely.
- Always turn off the camera before changing lenses.

13. Red-Eye Reduction/Self-Timer Lamp

That little circle on the front that lights up before a timed shot.

- Blinks when the self-timer is active
- Helps reduce red-eye in flash photography

Side Ports & Inputs

- **Left Side Flap:**
    - Micro USB (for transferring photos)
    - HDMI out (for connecting to screens)
- **Right Side:**
    - SD card compartment (accessed underneath with battery)

> Note: The Rebel T7 does **not** have an external microphone input. But don't worry—we'll talk about **audio workarounds** in the video chapter.

What to Press When Things Go Wrong

Here's your cheat sheet:

Problem	What to Do
Blurry Photo	Half-press shutter to lock focus, check shutter speed
Too Dark	Increase ISO or slow down shutter speed
Too Bright	Lower ISO or increase shutter speed
Wrong Colors (Too Blue/Orange)	Adjust White Balance or enable Auto WB
Not Focusing	Make sure lens switch is set to AF (not MF)
Shutter Won't Fire	Format your SD card, check battery, reinsert lens

These are common. You'll learn how to troubleshoot them fast.

The Hidden Button Combos You Should Know

Some of the Rebel T7's best features aren't obvious until you *combine* button presses.

- **Back-button focus trick** → Reassign focus to the "*" button (advanced tip in Chapter 8)
- **Exposure lock (AE Lock)** → Press "*" while metering to lock exposure
- **Zoom during playback** → Turn main dial or press "+"/"-" to zoom in and check sharpness
- **Live View toggle** → Press "Start/Stop" button to switch between screen and viewfinder

These save time and give you **pro-level control** with just a few presses.

Final Thought: Learn the Controls, Own the Camera

Most people use 20% of this camera's potential. Not because they're lazy—because no one showed them how it works in plain English.

Now you know:

- What every button does
- How to access critical features without digging through menus
- What to do when things go wrong
- Which shortcuts will save your shot

You're not fumbling anymore.
You're starting to *command* the camera.
And that's exactly what we want.

Chapter 4

Canon Menu Settings Explained Simply

(The No-Nonsense Guide to Setting Up Your T7 Like a Pro)

Your camera's menu system can feel like a maze the first time you open it. Tabs. Subtabs. Settings you didn't even know existed. But behind all that clutter are powerful tools that can make photography easier, cleaner, and faster—*if* you know what to turn on (and what to ignore).

This chapter breaks down

- Every essential Canon menu setting and what it *actually* does
- What you should change immediately—and why
- What to leave alone (at least for now)
- And how to create your own custom "quick-access" menu for speed

You don't need to learn everything.
You need to learn what matters.
So let's simplify.

Navigating the Canon Menu (How It's Structured)

Press the **MENU** button on the top-left of your Rebel T7. You'll see several **tabs**, divided into sections by icons:

- **Red Tabs**: Shooting Settings
- **Blue Tabs**: Playback Settings
- **Wrench (Yellow) Tabs**: Setup Settings
- **Camera + Star**: Custom "My Menu" (you can create this later)

You'll navigate through these tabs using

- The **Main Dial** (top scroll wheel) switch tabs
- The **Arrow Keys** (on the back) to move up/down
- The **SET button** (center) to select a setting
- The **Menu button** to go back one step

Let's Start with the Most Important Settings to Adjust First

Shooting Settings (Red Tabs)

This is where you configure how your camera takes and processes photos.

Image Quality

- **Set to: "L" (Large JPEG)** for now
- RAW vs. JPEG: Stick with JPEG if you're just starting. We'll explore RAW later.

Release Shutter Without Card → Disable

- Prevents the camera from pretending to shoot without a memory card.

Beep → Disable

- Turns off the annoying beeping noise every time you focus or press a button. Less distracting.

Grid Display → Enable

- Shows a helpful 3x3 grid on-screen while shooting. Great for composition and horizon lines.

Flash Control → Leave on E-TTL II

- Canon's automatic flash metering system. Adjusts flash brightness for you.
  (We'll go deeper into flash settings in Chapter 6.)

Playback Settings (Blue Tabs)

These settings control how your photos are reviewed.

Image Review → 2 sec

- Displays each shot for 2 seconds after taking it. Just long enough to catch a bad one.

Highlight Alert → Enable

- Shows flashing "blinkies" on overexposed areas during playback. Helps prevent blown-out skies or highlights.

AF Point Display → Enable

- Shows which autofocus point was used for each image when reviewing.

Setup Settings (Wrench/Yellow Tabs)

These are your general system preferences.

Auto Power Off → 1 min or 2 min

- Saves battery when idle. You can always wake it up with a half-press of the shutter.

LCD Brightness → Level 3 or 4

- Adjust to suit outdoor or indoor conditions.

Date/Time → Double-check accuracy

- Helps organize your photo files later.

Format Card → Do this now (if your card is new or hasn't been formatted before)

- Clears hidden data and prepares the card for optimal performance
  Remember: Formatting erases all files on the card.

Sensor Cleaning → Auto Clean on Startup/Shutdown

- Keeps your image sensor free of dust. Enable this and forget about it—Canon does the rest.

Create Your Custom "My Menu" Tab (Highly Recommended)

After a few weeks with the camera, you'll notice you keep adjusting the *same few settings*. Instead of digging through menus each time, let's create a **custom "My Menu" tab** where your favorite settings live.

To create it:

1. Navigate to the **green star tab** (My Menu)
2. Select **"Register to My Menu."**
3. Choose settings like
   - ISO setting
   - Format Card

- Image Quality
- Grid Display
- Auto Power Off
- White Balance

Now, instead of hunting through five different tabs, all your key settings are in one spot.

What You *Don't* Need to Touch Yet

- **Color Space**: Leave it on sRGB unless you're printing professionally
- **Long Exposure Noise Reduction**: Leave off for now
- **Custom Functions (C.Fn)**: Advanced tweaks—save for later chapters
- **Firmware Update**: Only do this when advised, using official Canon files

Hidden Menu Tips Most Users Never Discover

- **Live View Auto Focus**:
  In Live View, go to Menu > Live View Function Settings
  Set to **"FlexiZone"** for better control
- **Auto Lighting Optimizer**:
  Improves brightness in shadows.
  Set to **Standard** or **Low** for more natural results

- **Picture Style:**
  Choose "Standard" for now.
  But later chapters will show how to *customize this for your personal look*

- **Self-Timer Customization:**
  Set a 2-second or 10-second delay with multiple shots. Great for group selfies.

Real-World Shortcut: Quick Control Menu ("Q" Button)

The fastest way to change settings *without diving into the full menu* is to press the **"Q" button** while in shooting mode.

From there, you can:

- Change ISO
- Adjust White Balance
- Choose Autofocus Mode
- Switch Drive Modes (single shot vs burst)
- Set Picture Style
- Apply Exposure Compensation

Pro Tip: Learn to love the Q button. It's the fastest route to controlling your camera on the fly.

Final Recap: Your Menu, Your Rules

By now, you've done what 80% of Rebel T7 owners never do:

- You've learned the *menu*, not just the buttons
- You've customized your setup for smarter shooting
- You've disabled the annoying stuff and turned on the helpful stuff
- And you've created a shortcut menu that actually works the way *you* shoot

Photography shouldn't feel like hacking into a machine. It should feel like second nature. This chapter just moved you one giant step closer.

Chapter 5

Shooting Modes Demystified

When to Use Auto, Manual, and Everything In Between

This is where your Canon EOS Rebel T7 starts to come alive.

Because now, it's no longer just a device—it's a partner. And knowing how to control that partnership comes down to this: **understanding your camera's shooting modes.**

Each mode on the dial isn't just a setting—it's a *philosophy* for how your camera behaves.

Do you want to freeze motion? Control the background blur? Let the camera handle most decisions, or take full control yourself? The right mode makes that possible.

In this chapter, you'll discover:

- What every mode on the Rebel T7's dial really means (without the fluff)
- When and *why* to use each mode
- The traps beginners fall into—and how to avoid them
- Hidden strengths of overlooked modes

- A simple cheat system for choosing the best mode in any scenario

This is your bridge from taking random photos to **making intentional images.**

Let's decode the dial.

Meet the Mode Dial

On top of your Canon Rebel T7 is the **Mode Dial**—that round wheel with letters and symbols. It's the brainstem of your camera. It tells the T7 *how much control* you want.

Canon split the modes into two categories:

Basic Zone (Auto & Presets)

- For people who want fast results without thinking much

Creative Zone (Manual & Priority Modes)

- For those ready to take creative control

Let's start simple and work up.

BASIC ZONE MODES (Beginner-Friendly, But Limiting)

These modes automate most decisions.

Scene Intelligent Auto (Green Box Icon)

- *"I just want the camera to do everything for me."*
- Great for: total beginners, quick snapshots, fast-paced moments

The camera chooses everything—shutter speed, aperture, ISO, white balance, etc.

> But beware: you sacrifice creative control. You can't change settings like ISO or white whitebalance manually. Also, the flash may pop up even when you don't want it.

Creative Auto (CA)

- *Auto, with training wheels for creativity*
- Lets you change background blur, flash, drive mode, and color tone using simple sliders

Good for beginners wanting to *ease into manual decisions* without jumping to full manual mode.

Portrait Mode (Icon: Face)

- Prioritizes a blurred background (shallow depth of field)
- Softens skin tones and enhances color warmth

Great for headshots, pets, and social photography.

Landscape Mode (Mountains Icon)

- Sharpens the entire scene (deep focus)
- Boosts greens and blues for dramatic skies and nature shots

Ideal for outdoor scenes—but useless indoors.

Sports Mode (Running Person Icon)

- Uses fast shutter speed to freeze motion
- Prioritizes continuous autofocus (AI Servo)

Perfect for moving subjects: kids, pets, dancers, and athletes.

Night Portrait Mode (Person + Star Icon)

- Combines longer shutter speed + flash
- Tries to capture background ambient light *and* foreground subject

Use a tripod—this mode slows the shutter and risks blur.

When to Use Basic Zone:

- You're brand new to photography
- You're handing the camera to a friend or family member
- You're in a rush and need a decent shot now
- You don't have time (or light) to adjust settings

But remember: You're trading creativity for convenience.

CREATIVE ZONE MODES (Where the Magic Happens)

This is where you grow. These modes give *you* control—fully or partially.

Program Mode (P)

- The camera sets shutter and aperture, but *you* can tweak ISO, white balance, and exposure compensation

Think of it as "Auto with Options."
Great for: quick shooting with flexibility

Shutter Priority (Tv Mode)

- *You set the shutter speed*, camera picks aperture
- Use this when motion control is essential

Examples:

- 1/1000 sec → Freeze sports, water splashes
- 1/60 sec → Standard handheld portrait
- 1/15 sec → Light trails, motion blur

***Tip:** If your photos turn out too dark in TV mode, it's because your camera can't open the lens wide enough. Try raising ISO or using a brighter scene.*

Aperture Priority (Av Mode)

- *You set the aperture (f-stop)*, camera chooses shutter speed
- Use this to control **depth of field** (how blurry the background is)

Examples:

- f/2.8 → Blurry background for portraits
- f/8 → Balanced landscape
- f/16 → Maximum sharpness for architecture

This is the go-to mode for many photographers.

Manual Mode (M)

- You control *everything*: shutter speed, aperture, ISO, white balance, etc.

Sounds scary? It's not.

Use manual mode when:

- Lighting is stable (e.g., studio or outdoors in steady sun)
- You want consistent exposure across multiple shots
- You're experimenting or learning how each setting affects the image

Secret: Manual Mode is *often easier* than fiddling with Auto—because you're in control.

Bonus Mode: Exposure Compensation (+/- Button)

Not a mode, but an essential creative tool.

If your photo is too dark or too bright, you don't need to switch modes—just use the **Exposure Compensation button (+/-)** and turn the dial to brighten or darken the image in real-time.

It's the fastest way to fix exposure without touching ISO or aperture.

Which Mode Should You Use? A Quick Guide

Situation	Best Mode
Fast-moving subjects	Tv (Shutter Priority) or Sports Mode
Portraits with blurred background	Av (Aperture Priority)
Bright day, detailed shots	Av or Landscape
Indoors, low light	M or Av (with higher ISO)
Casual snapshots	Auto or P Mode
Total control needed	Manual (M)
Artistic experimentation	Manual (M) or Av

Pro Tips for Shooting Mode Mastery

- **Use Auto to get the shot.** Use the manual to *understand* the shot.
- **Switch modes as light or subject changes.** One size doesn't fit all.
- **TV and AV are powerful training modes.** They let you control one creative element without overloading your

brain.

- **Manual mode isn't "advanced."** It's just deliberate. You already have what it takes.

Final Word: Modes Are Just Tools—*You* Are the Artist

Your Rebel T7 has the power to help you see the world differently. But it's not about picking the "right" mode. It's about picking the *right mode for you*—for what you want to say with your image.

Think of each mode as a lens into a different way of shooting:

- Want emotion? Use shallow depth.
- Want drama? Freeze motion or blur it with intent.
- Want a story? Use light, contrast, and focus to direct the viewer's eye.

Your camera listens to you through the mode dial.
And now—you finally know what to say.

Chapter 6

Camera Settings You Need to Master

You've set up your Canon EOS Rebel T7. You understand the buttons. You know which shooting modes to use. Now comes the part that separates average shooters from confident creators: **understanding the core camera settings that shape every photo.**

But we're not doing this the Canon way.
We're doing this the *human* way.

No graphs. No math. No dense technical jargon.
Just clear, experience-driven insight into the essential settings—what they do, why they matter, and how to use them *without getting stuck.*

This chapter covers:

- The real meaning of the Exposure Triangle (ISO, Shutter Speed, Aperture)
- Autofocus modes, image quality, and white balance
- How to confidently adjust settings in the field without second-guessing
- What settings to master now—and which ones can wait

If you've ever taken a photo that came out **blurry, too dark, too bright**, or **weirdly colored**... This chapter is about to fix that.

The Exposure Triangle — Made Stupid-Simple

You may have heard this term. It sounds fancy. But it's just this:

Three settings that control how light enters your camera and affects your image. Change one, and you affect the others.

Let's break it down.

1. ISO — How Sensitive Your Camera Is to Light

- Low ISO (100–400) = Crisp, clean images in bright light
- High ISO (800–6400+) = Brighter images in low light *but more grain*

Think of ISO like your eyes:

- During the day, you don't need to strain (Low ISO)
- At night, you open your eyes wide to see—but things get fuzzy (High ISO)

Tip: Always start at ISO 100 in bright scenes. Only raise it when light gets tricky.

2. Shutter Speed — How Long Your Camera's Eye Stays Open

Measured in fractions of a second:

- 1/1000 → Very fast → Freezes motion (sports, action)
- 1/125 → Mid-range → Good for portraits and daily use
- 1/30 → Slower → Risks blur without tripod
- 1" → One full second → Used for night scenes, light trails

Rule of Thumb:

- Use **1/60 or faster** when handheld to avoid blur.
- Slower = more light, but also more motion blur.

3. Aperture (f-stop Wide the Lens Opens

- f/1.8 = Wide opening = More light = **Blurry background**
- f/5.6 = Medium depth = Good for general shots
- f/11 or f/16 = Narrow = Less light = **Everything sharp**

Aperture is how you **control depth of field:**

- Want dreamy portraits? Use a low f-number.

- Want everything in focus (like landscapes)? Use a higher f-number.

How They Work Together

- **Dark Photo?** Raise ISO or use slower shutter speed
- **Too Bright?** Lower ISO, faster shutter, or smaller aperture
- **Blurry Subject?** Use faster shutter or raise ISO
- **Flat Image?** Try wide aperture (f/2.8) for depth

You'll eventually *feel* the balance of these three. But for now, experiment. Adjust one at a time. Observe what changes.

Autofocus (Getting Sharp, Every Time

Press the **AF** button or go through the **Q menu** to find your autofocus settings.

Canon T7 Focus Modes:

- **One-Shot AF**: Best for still subjects (portraits, landscapes)
- **AI Servo AF**: Tracks moving subjects (sports, pets, kids)
- **AI Focus AF**: Auto-switches between One-Shot and AI Servo
(works, but not always reliable—best to pick yourself)

Most people leave it on One-Shot—and miss moving subjects.

Focus Point Selection

By default, your camera chooses focus points. That's risky. It may lock onto a background tree instead of your subject's eyes.

Press the **AF Point Selection button** (top-right of the back panel):

- Choose **a single point** and move it manually with arrow keys.
- Place it on what you want sharp—like someone's eyes.

White Balance (Get Natural-Looking Colors

White balance tells your camera what "white" should look like under different lighting.

Modes:

- **Auto White Balance (AWB)** → Good most of the time
- **Daylight** → Clear sun
- **Cloudy** → Warmer tones
- **Shade** → Very warm
- **Tungsten/Incandescent** → Corrects yellow indoor lighting
- **Fluorescent** → Cool indoor tones
- **Custom** → Advanced (we'll cover this later)

Tip: Use "Cloudy" to warm up outdoor photos—it flatters skin tones.

Picture Style — Your Camera's "Mood Filter"

Found in the **Q menu** or **shooting settings.**

- **Standard** → Balanced contrast and color
- **Portrait** → Softer skin tones
- **Landscape** → Boosts blues and greens
- **Neutral** → Low contrast, great for editing
- **Monochrome** → Black and white images

Stick with **Standard** for now. Once you get comfortable, start experimenting.

Image Quality — What File Type Should You Use?

- **L (Large JPEG)** = High-quality, small file, easy to use
- **RAW** = Uncompressed, best quality, flexible for editing, but large files
- **RAW + JPEG** = Both versions (uses more space)

For now, use Large JPEG.
Once you're ready to edit and shoot professionally, switch to RAW.

Real-World Settings for Common Scenarios

Scene	ISO	Shutter Speed	Aperture	Mode
Bright Day Portrait	100	1/200	f/2.8	Av
Indoor Without Flash	800–1600	1/60	f/3.5	M or Av
Sports/Action Outdoors	400	1/1000	f/5.6	Tv or M
Landscape at Sunset	100	1/60 or slower	f/11	Av
Night City Scene	1600+	1/10	f/4.5	M (tripod)

Adjust from here based on light and your desired look. These aren't rules—they're starting points.

The Five Settings You Should Master First

If nothing else, get fluent in these:

1. **ISO** — Always check it before shooting
2. **Shutter Speed** — Especially for moving subjects
3. **Aperture** — Controls blur and focus depth
4. **AF Mode + Point Selection** — Critical for sharp subjects
5. **White Balance** — For natural skin tones and colors

Get these right, and your photos will *consistently* look intentional, not accidental.

Final Words: Settings Are Just a Language

This chapter might feel like a lot. That's okay.

You’re not expected to memorize every number. You’re learning a new *language*—one that lets your camera listen to your creative choices.

And with practice, the dial becomes instinct. The ISO, the shutter, the focus—they stop being numbers and start becoming expressions of how *you* see.

You're not just pressing buttons anymore.
You're composing. You're communicating.
You're creating.

Chapter 7

Manual Mode Made Simple

Let's tell the truth nobody else does: **Manual Mode isn't scary.**

It's not advanced. It's not for pros only. It's just you, telling your camera exactly what to do—nothing more.

In fact, once you understand the few moving pieces involved, manual mode becomes the most *liberating* way to shoot. No more guessing. No more hoping the camera "gets it right." You're not reacting—you're *leading.*

This chapter is your on-ramp to creative control.

We'll cover:

- A clear, hands-on walkthrough for using Manual Mode for the first time
- How to balance ISO, aperture, and shutter speed—without getting stuck
- When to use Manual over Auto or Av/Tv
- Hidden techniques to make Manual Mode faster and more intuitive

- Confidence boosters to take the fear out of full control

So take a breath. Forget the word "manual" for a second. You're just telling your camera how you want your photo to look. That's it.

Why Manual Mode Matters

Here's what Manual Mode gives you that Auto never will:

- Consistent exposure across all your photos
- Full control over mood and depth
- Precision for tricky lighting (backlight, concerts, night)
- Mastery over motion—freeze or blur it exactly how you want
- A creative voice that says, *"I decided to shoot this way on purpose."*

And most importantly:

It teaches you to see light, not just capture it.

The 3 Settings You Control in Manual Mode

When you switch your mode dial to **M**, three things become your responsibility:

1. **Aperture** (f-Controls how much light enters *and* how blurry your background is)
2. **Shutter Speed** — Controls how long the light hits the

sensor (and whether your image is sharp or blurred)

3. **ISO** — Controls how sensitive your camera is to light

These three settings form the exposure triangle we covered in Chapter 6.

You control them with:

- **Main Dial (Top right wheel)** → Usually controls shutter speed
- **Press the "Av" (+/-) button and turn dial** → Adjust aperture
- **Use ISO button on top or Q menu** → Change ISO

Now let's use them in real time.

Your First Manual Photo (Step-by-Step)

Let's set up a basic indoor portrait in natural light.

1. **Switch to Manual Mode (M)** on the Mode Dial
2. **Set ISO** to **400** (if indoors with daylight coming through a window)
3. **Set Aperture** to **f/4.0** (gives some background blur but keeps the subject clear)
4. **Set Shutter Speed** to **1/125** (safe for handheld, no motion blur)

That's your **light meter.** It tells you how bright your exposure is.

- **0** = balanced exposure
- **-1 to -2** = underexposed (too dark)
- **+1 to +2** = overexposed (too bright)

Adjust one setting at a time until the meter hovers close to 0.

Try it now.

- Is it too dark? Either slow the shutter speed (e.g., 1/80), open the aperture (e.g., f/2.8), or raise the ISO (e.g., 800).
- Too bright? Do the opposite.

Take the photo.
Check the result.
You just shot in full manual.

Common Manual Mode Situations (And What to Set)

Here's how you can approach different scenarios using Manual Mode:

Outdoor Portrait in Bright Light

- ISO: 100
- Aperture: f/2.8 (blurry
- Shutter: 1/1000 or faster (to avoid overexposure)

City Street at Night

- ISO: 1600+
- Aperture: f/3.5
- Shutter: 1/30 (use a tripod if slower)

Sports or Fast Action

- ISO: 400
- Aperture: f/5.6
- Shutter: 1/1000

Creative Blur (Light Trails, Waterfalls)

- ISO: 100
- Aperture: f/11
- Shutter: 1/4 or slower (use tripod)

Pro Tip: Shutter speed affects motion, aperture affects depth, and ISO affects clarity. Pick what matters most for the shot—and adjust the rest to support it.

Hidden Tricks for Making Manual Mode Easy

Use Exposure Simulation (Live View)

- Press **Start/Stop** button to enable Live View

- Your LCD now shows real-time brightness as you adjust settings
- It's like having a live exposure preview—*use it often*

Enable the Light Meter in Viewfinder

- Watch the [-2...0...+2] scale while adjusting shutter/aperture
- It becomes your exposure compass

Create a "Safe Starting Point" Preset

Save one set of reliable manual settings you can return to when you're stuck.

- ISO 400
- Aperture f/5.6
- Shutter 1/125

It works in most lighting and gives you a foundation to tweak from.

"What If I Mess Up My Settings?"

Great. You're learning.

Manual mode invites trial and error. Every "mistake" becomes a mini-lesson in light, motion, and mood. That's how every great photographer got started.

Photography isn't about getting everything perfect. It's about being present, observing, and responding.

You'll begin to know *instinctively* when a scene needs a brighter exposure or a softer background. Because now, you're not waiting on the camera. You're calling the shots.

Final Word: Manual Mode is a Conversation with Light

Each photo in Manual Mode is a quiet conversation between you and the moment. It asks:

- What do I want to say here?
- What emotion should this frame hold?
- What do I want the viewer to *feel*?

The camera stops being a black box. It becomes a translator—for your perspective, your emotion, and your story.

And you?

You're not just shooting anymore.

You're *directing* the light.

Chapter 8

Focus, Sharpness & Color That Pops

A photo that isn't sharp is a photo that gets deleted. A portrait where the eyes are out of focus? Gone. A beautiful moment captured in dull, lifeless color? Skipped.

This chapter is about fixing that forever.

Because getting your camera *technically* set up is only half the equation. The other half—the one that makes a photo look professional—is **how you direct focus, lock in sharpness, and pull out natural color** that sings on the screen.

The Canon Rebel T7 is capable of stunning image quality. You don't need a more expensive lens. You don't need editing tricks. You just need to unlock a few tools your camera already has—but rarely explains well.

Let's dive in.

The Power of Focus (And Why Most People Let the Camera Guess Wrong)

Here's what most new users don't realize:

By default, your Canon T7 chooses where to focus—and it often *guesses guesseswrong.*

It might lock onto:

- A tree in the background
- A person's shirt instead of their eyes
- A random contrast line instead of your subject's face

That's why *you* need to tell it exactly where to focus.

Step 1: Choose the Right Autofocus Mode

Your Rebel T7 has three primary autofocus (AF) modes:

1. **One-Shot AF** — For still subjects. Once it locks focus, it holds it.
    - Use for portraits, landscapes, still life, and products.
2. **AI Servo AF** — For moving subjects. Continuously tracks motion and adjusts.
    - Use for sports, kids, pets, cars, dancing, or walking people.
3. **AI Focus AF—The camera** decides whether to use One-Shot or AI Servo.
    - Sounds smart. Often it isn't. Avoid for now.

⊡ **Recommendation: Start with One-Shot AF. Switch to AI Servo AF when your subject moves *toward or away from you.***

You can change your AF mode via:

- **Q Button > AF Mode**
- Or: Press **down arrow** on back dial shortcut

Step 2: Take Control with Focus Point Selection

This is the single biggest game-changer for beginners.

When you half-press the shutter button, where your camera focuses depends on which *focus point is active.*

By default, the camera chooses this for you. You're going to change that.

Here's how:

1. **Press the AF Point Selection button** (top-right, near the viewfinder)
2. Use the **arrow keys** to move the active point to the spot you want
3. Place the focus point directly over your subject—*especially their eyes*

⊡ **Pro Tip**: In portraits, always focus on the **eye closest to the camera.** That's what viewers connect to first.

Step 3: Activate the "Back-Button Focus" Trick (Optional Power Move)

This technique separates **focus** from **shutter**, giving you more precision.

Instead of half-pressing the shutter to focus, you'll reassign focusing to the **"*" (star) button** on the back of the camera.

This lets you:

- Lock focus with your thumb
- Recompose your shot
- Then take the photo *without losing focus*

It's cleaner. Sharper. More accurate.

> ⊡ Want to enable this? It takes a few menu changes.
> I'll walk you through this in Chapter 11 (Features & Hacks)

Now, Let's Talk Sharpness

Autofocus gets you close. But sharpness is a **combination of factors:**

1. Shutter Speed Matters

If your shutter is too slow, even a steady hand will create blur.

Lens Length	Minimum Safe Shutter
18mm (wide)	1/30 sec
55mm (zoomed in)	1/60 or 1/125 sec
Moving subjects	1/500 – 1/1000 sec

Rule of thumb: Use a shutter speed **at least 1/focal length** (e.g., 1/50 for a 50mm lens).

2. Hold Your Camera Like a Pro

- Right hand grips the camera firmly
- Left hand **supports the lens from below** (not the side)
- Elbows close to your torso
- Feet shoulder-width apart
- Lean against a wall or object if needed for extra stability

The way you *hold* the camera is one of the most overlooked factors in sharpness.

3. Enable Image Stabilization (IS)

If your lens has a switch labeled **"IS,"** make sure it's **ON**.
This helps reduce blur from small hand movements—especially in low light.

Color That Pops: Get Natural, Lifelike Results

Now let's shift from sharpness to **vibrancy**.

Bad color happens when your camera misjudges the lighting. The solution?
White balance and picture style.

White Balance = Color Temperature Control

You already know the basics from Chapter 6.

Here's the takeaway:

- **Auto White Balance (AWB)**: Good starting point
- **Cloudy Mode**: Adds warmth—great for skin tones
- **Shade Mode**: Even warmer, for sunrise/sunset
- **Custom White Balance**: Use a gray card for total control

□ *Trick:* In golden hour (sunset), use **Cloudy** to enhance the warm glow.

Picture Styles = Your Camera's Personality

Think of Picture Styles as *Instagram filters—baked into the photo.*

Choose yours by:

- Pressing the **Q button**
- Navigating to **Picture Style**

Options:

- **Standard** → Balanced look
- **Portrait** → Softer tones, flatter skin

- **Landscape** → Vivid blues and greens
- **Neutral** → Flat tones for editing
- **Monochrome** → Black & white shooting

Try Portrait for skin, Landscape for outdoor greens, and Standard for all-purpose.

Summary Checklist: How to Get Pro Results Every Time

- ☐ Use **One-Shot AF** for stills and **AI Servo AF** for motion
- ☐ Manually select your **focus point**—don't let the camera guess
- ☐ Keep **shutter speed fast enough** for sharp handheld shots
- ☐ Support the camera correctly—form matters
- ☐ Set **White Balance** based on lighting conditions
- ☐ Use **Picture Style** to match your scene or mood
- ☐ Zoom in during playback to **check focus on eyes**

☐ Final Thought: Great Images Aren't Accidental

They're designed. Directed. Created.

Getting sharp, vibrant, emotionally resonant images isn't about having better gear—it's about being deliberate with your focus and your light.

With these tools, you're not just capturing what's there.
You're *revealing* it.

The difference between a snapshot and a photograph?
Intent.
And you now have it.

Chapter 9

Video Recording with the Rebel T7

Still photos freeze time.
Video lets you *feel* it move.

Whether you're capturing a child's laughter, a street performer's rhythm, a behind-the-scenes vlog, or a slow-motion memory, your Canon EOS Rebel T7 is more than capable of bringing your stories to life with motion. But there's a catch…

> **The T7 *wasn't built for video pros.* Yet, with a few smart adjustments, it can shoot video that looks professional—even without extra gear.**

This chapter is your no-fluff, real-world guide to:

- Shooting beautiful, steady, clean-looking video on the T7
- Avoiding the common mistakes that make beginner videos unusable
- Understanding frame rate, focus, and audio limitations
- Unlocking tricks and workarounds to go beyond the camera's built-in restrictions

And remember:
Video isn't just about gear. It's about *feel.*
If you can tell a story with light, pacing, and presence, you're already halfway there.

What Your Rebel T7 *Can* Do With Video

Before we shoot, know what's under the hood:

- **Full HD 1080p video resolution** (no 4K)
- **24fps or 30fps frame rate** (no 60fps slow motion)
- **Live View screen for framing**
- **Manual exposure and focus available**
- **No external mic input** (we'll fix that later)
- **Basic video compression (MOV format)**

What this means: You can absolutely shoot crisp, cinematic-looking video—with limitations. But every limitation has a workaround, which we'll cover below.

Step-by-Step: How to Shoot Your First Video

1. Switch to Movie Mode

- Turn the **Mode Dial** to the video camera icon
- Flip the **Live View switch** (Start/Stop button) to activate video mode

You'll see the LCD screen activate with a real-time preview. This is your viewfinder now.

2. Set Your Video Settings

Press **Menu** and scroll to the **Shooting Settings** tab for video.

Recommended Settings:

- **Movie Recording Size** → 1920x1080, 30fps or 24fps
    - 30 fps = smooth motion (vlogging, daily life)
    - 24 fps = cinematic (narratives, interviews)
- **Sound Recording** → Auto (or disable if using external audio)
- **Auto Lighting Optimizer** → Disable (gives more natural contrast)
- **Picture Style** → Neutral or Portrait for skin tones

3. Set Your Exposure Manually *(Highly Recommended)*

Use **Manual Mode (M)** for video exposure.
Auto exposure causes constant brightness shifts—*not good.*

Suggested starting point (outdoors):

- ISO: 100
- Shutter Speed: 1/60
- Aperture: f/4.5

- White Balance: Daylight or Cloudy (never AWB)

Rule of thumb for shutter speed:
Always set it to **double your frame rate** (e.g., 1/60 for 30 30fps).

4. Set Your Focus

The Rebel T7 **does not have Dual Pixel autofocus** like newer Canon models. That means its continuous autofocus during video is **slow and unreliable.**

Instead:

- **Use manual focus.**
  - Flip the lens switch to **MF**
  - Rotate the focus ring while watching the screen
  - Zoom in with the **magnify button (+)** to check sharpness
  - Lock it in before you record

This gives you the sharpest and most consistent results.

5. Press Record (Start/Stop Button)

Once you're happy with framing, exposure, and focus, hit the **Start/Stop button** to begin recording.

A red dot will appear on-screen.

Audio: The Rebel T7's Weakest Link (And How to Fix It)

Your T7 does **not** have an external mic input.
That's a serious limitation. The built-in mic is

- Directionless
- Prone to wind noise
- Picks up every touch and button press

Solution: Use a Separate Audio Recorder

Record clean sound using:

- Your **smartphone with a lav mic** (cheap and surprisingly good)
- **A Zoom H1n or Tascam recorder**
- A wireless mic system like **Rode Wireless GO (record to phone or second camera)**

Then **sync the audio and video** later using free software like DaVinci Resolve or iMovie.

Pro Hack: Clap once at the start of your video to help sync the audio visually (spike in waveform) and by sound.

Tips for Smoother, Cleaner Video

⊡ Use a Tripod or Stabilizer

Even the steadiest hands introduce a shake. Use:

- A **tripod** for static shots
- A **gorillapod** for handheld walkarounds
- A **cheap gimbal** for cinematic movement (optional)

⊡ Nail Your White Balance Before Recording

Video is hard to correct later. Always set WB *before* you hit record.

⊡ Set Picture Style to "Neutral."

Gives you more control in editing. No baked-in contrast or over-sharpening.

⊡ Use Manual Focus for Consistency

Even if it takes longer, it avoids that embarrassing mid-record focus hunt.

⊡ Shoot in Short Takes

Don't record long, continuous clips. Instead:

- Break your video into 10–30 second takes
- Helps you stay organized and focused
- Easier to edit, rearrange, and sync later

Advanced Tricks for Getting More Out of the T7

☐ Use Exposure Simulation in Live View

Turn it on via:
Menu > Shooting Settings > Expo. Simulation → Enable

This shows real-time changes to your exposure on-screen.

☐ Enable Grid Display for Framing

Helps align horizons, center subjects, and follow the rule of thirds.

Menu > Shooting Settings > Grid Display → Enable

☐ Use "Movie Servo AF" Only if Subject Isn't Moving Much

If you absolutely must use autofocus during recording:

- Set to **Movie Servo AF: Enable**
- Use in *bright light* only
- Lock exposure manually to avoid flickering

A Sample Setup: Talking Head Video Indoors

Setting	Value
Mode	Manual (M)
ISO	400
Shutter Speed	1/60
Aperture	f/4
Frame Rate	1920x1080 @ 30fps
Focus	Manual Focus on eyes
White Balance	Custom or Daylight
Tripod	Yes

After You Record: How to Get the Best Results

- **Transfer video via USB or SD card reader**
- **Use free video editors**: iMovie (Mac), DaVinci Resolve (Windows/Mac), or VN app (mobile)
- **Sync audio manually** if recorded separately
- **Add music or light color correction** for polish

Great storytelling always beats perfect settings. But now, you'll have both.

Final Word: Your Camera Tells Stories in Motion, Too

Even with its limitations, the Rebel T7 is more than enough to document life, emotion, and memory in motion.

Just remember:

- Audio matters as much as visuals

- Focus is best when locked before recording
- You don't need flashy transitions—just presence and intention

A single steady clip of someone laughing.
A simple street scene.
A quiet moment in good light.
These are the things that endure.

Your camera's rolling.
Time to direct the moment.

Chapter 10

Real-World Shooting Recipes & Scenarios

So far, we've walked through theory, controls, modes, and manual mastery.
Now it's time to make it real.

This chapter is your go-to field guide—no fluff, no guesswork. We'll walk through **specific real-life shooting scenarios** and give you:

- The exact **settings that work**
- What **to watch out for**
- Pro-level **composition and light tips**
- And ways to build confidence every time you shoot

These "shooting recipes" are tailored for **you—the beginner with purpose.**
Whether you're at a family gathering, a hiking trail, a night market, or just your backyard, this is how you start taking photos that *feel* intentional.

1. **Portraits — Capturing People With Depth**

Goal: Sharp subject, soft background, flattering skin tones

Settings:

- **Mode:** Av (Aperture Priority)
- **Aperture:** f/2.8 – f/4.5
- **ISO:** 100 (outdoors), 400–800 (indoors)
- **Shutter Speed:** Auto (but try to keep above 1/125)
- **White Balance:** Cloudy (adds warmth)
- **AF Mode:** One-Shot
- **Focus Point:** Manually select and place on the closest eye

Tips:

- Position subject facing soft, natural light (window, open shade)
- Avoid harsh noon light—early morning or golden hour is best
- Use a plain background or one with distance for best blur
- Ask your subject to slightly turn their body and keep their eyes on the camera

⊡ ***Make it better:*** *Add a subtle catchlight in the eyes by facing them toward light.*

2. Landscapes — Wide, Sharp, and Vivid

Goal: Everything in focus; rich skies and color

Settings:

- **Mode:** Av
- **Aperture:** f/8 – f/11
- **ISO:** 100
- **Shutter Speed:** Auto (tripod if slower than 1/60)
- **White Balance:** Daylight or Shade
- **Picture Style:** Landscape
- **Focus Point:** Middle third of the scene

Tips:

- Use a tripod for razor-sharp results, especially in low light
- Compose with a strong foreground (rocks, plants, water)
- Try sunrise/sunset for dramatic skies
- Turn on the grid display and align with the rule of thirds

□ *Make it better: Use a polarizing filter to deepen skies and reduce reflections (optional).*

3. Action & Movement — Freeze the Moment

Goal: Sharp moving subjects (sports, pets, kids)

Settings:

- **Mode:** Tv (Shutter Priority)
- **Shutter Speed:** 1/500–1/1000
- **ISO:** 400–800
- **Aperture:** Auto
- **AF Mode:** AI Servo (continuous focus)
- **Drive Mode:** High-speed continuous
- **Focus Point:** Center or zone focus

Tips:

- Follow your subject with the camera (panning)
- Shoot in burst mode to increase your chances
- Focus just ahead of moving subjects for predictive sharpness

□ ***Make it better: Anticipate the action. Don't react—predict.***

4. **Sunset & Golden Hour — Warmth and Mood**

Goal: Capture the mood of late-day light with rich color and soft shadows

Settings:

- **Mode:** Av
- **Aperture:** f/5.6 – f/8
- **ISO:** 100–200
- **White Balance:** Cloudy or Shade
- **Exposure Compensation:** -0.3 to -1.0 (to preserve highlight color)

Tips:

- Include silhouettes or strong foreground shapes
- Face the camera slightly away from the sun to avoid lens flare
- Use spot metering if your subject is in the shadow

Make it better: Shoot both wide and tight—details can tell as much story as the sky.

5. Night Scenes & Low Light – Without a Flash

Goal: Clean, sharp images at night without using harsh flash

Settings:

- **Mode:** M (Manual)

- **Shutter Speed:** 1/10–10 sec (use tripod)
- **Aperture:** f/4 – f/8
- **ISO:** 800–1600
- **White Balance:** Tungsten or custom
- **Picture Style:** Neutral

Tips:

- Always use a tripod or stable surface
- Use the 2-second self-timer to avoid shake
- Focus manually or use Live View magnification for clarity

□ *Make it better: Underexpose slightly to preserve city lights and color.*

6. Street & Travel Photography — Fast, Flexible, Discreet

Goal: Capture life as it happens without fuss or bulk

Settings:

- **Mode:** P (Program Auto) or Av
- **ISO:** Auto (limit to 1600)
- **Aperture:** f/5.6 – f/8

- **White Balance:** Auto
- **AF Mode:** One-Shot or AI Focus
- **Drive Mode:** Single shot or silent
- **Picture Style:** Standard or Portrait

Tips:

- Keep your camera turned on and settings dialed in
- Frame through the viewfinder for subtlety
- Use shadows, reflections, or juxtapositions to tell stories

□*Make it better: Look for human moments, not just buildings.*

7. **Food & Product Photography — Small Scenes, Big Impact**

Goal: Bright, sharp, appetizing images

Settings:

- **Mode:** Av
- **Aperture:** f/2.8 – f/5.6
- **ISO:** 100–400
- **White Balance:** Custom or Daylight
- **AF Mode:** One-Shot
- **Focus Point:** Manually place on the focal item (e.g.,

garnish)

Tips:

- Use natural side lighting from a window
- Add a bounce card (white paper) to soften shadows
- Use a tripod or surface for sharper detail

□*Make it better: Add human elements—a hand, a fork, a spill—for story.*

How to Build Confidence Through Practice

Every scenario above is a chance to **learn in public.**
Real light. Real moments. Real variables.
You're not just adjusting a camera—you're responding to life.

Do this:

- Pick one shooting recipe per week
- Challenge yourself to shoot 10 photos in that style
- Evaluate what worked, what didn't, and why

Repeat weekly.
You'll build intuition faster than reading any manual.

Final Word: Photography Is an Interaction

It's not just about exposure or sharpness.
It's about paying attention—to light, to story, to connection.

These scenarios aren't just technical guides.
They're *starting points for presence.*

You now have the tools to shoot with purpose.
And soon, your photos will start to speak—clearly, confidently, and beautifully.

Chapter 11

HIDDEN FEATURES, TRICKS & WORKAROUNDS

By now, you've mastered the essentials: camera setup, shooting modes, manual control, and real-world photography recipes.
But the Canon EOS Rebel T7 still has a few **hidden tools, clever shortcuts, and workarounds** that Canon won't highlight in the official manual—but *you need to know.*

This chapter is where you discover

- Power-user tricks buried in the menus
- Little-known functions that make your workflow smoother
- Secret settings that help you get cleaner shots
- Creative "hacks" that unlock surprising results with basic gear
- And workarounds for the T7's biggest weaknesses (like no mic jack)

These aren't gimmicks. These are **field-tested enhancements** to your shooting experience—ones that separate casual users from confident creators.

Let's dig in.

1. Back-Button Focus — More Control, Less Confusion

Default behavior: Pressing the shutter button halfway activates focus. Problem: This causes focus to shift at the worst possible time—just before you take the photo.

Back-button focus separates your *focus* from your *shutter*. You focus with your thumb (the "*" **button**) and shoot with your index finger—like a pro.

Why it matters:

- Lets you lock focus and recompose freely
- Prevents the camera from re-focusing accidentally
- Makes manual adjustments faster and more consistent

How to Set It Up:

1. **Menu > Custom Functions (C.Fn) > C.Fn-9: Shutter/AE Lock Button**
2. Change to: **1: AE Lock / AF**

Now:

- Press the "*" button (top-right rear) to focus
- Press the shutter button fully to shoot

> **Tip: This is especially helpful for portraits, off-center subjects, and street photography.**

2. Exposure Simulation — Real-Time Brightness Preview

This is a **silent game-changer** when shooting in manual mode.

When enabled, your LCD screen shows the actual brightness of your image as you adjust settings. No surprises.

How to Enable:

- **Menu > Live View Shooting Settings > Expo. Simulation → Enable**

Now when using **Live View**, you can literally watch the screen get darker or brighter as you change shutter speed, aperture, or ISO.

3. Sensor Dust Mapping—Built-In Protection for Your Images

Your T7 includes an **internal dust detection system** to help you manage sensor cleanliness.

Dust on the sensor appears as small spots on your photos—especially in skies or solid backgrounds.

How to Use:

- **Menu > Sensor Cleaning > Clean Now or Clean at Shutdown**
- Optionally: **Dust Delete Data** to map dust spots for editing later

You still need to physically clean your sensor sometimes, but this auto-cleaning helps delay that need.

4. Picture Style Customization—Create Your Own "Look"

Picture Styles control sharpness, contrast, color tone, and saturation. But did you know you can **customize** them?

How to Create a Custom Style:

1. **Menu > Picture Style > Select "User Def. 1"**
2. Press **INFO** to adjust:
    - Sharpness
    - Contrast
    - Saturation
    - Color tone

Create a soft look for portraits, a high-contrast style for street shots, or a muted style for editing later.

Try reducing contrast and sharpness slightly for a flatter, more cinematic image—great for post-processing.

5. Digital Zoom in Playback — Check Focus with Precision

You can zoom in *after taking a shot* to inspect sharpness or focus accuracy.

- Press the **magnify (+)** button while reviewing an image
- Use the **arrow keys** to pan around
- Press **magnify (-)** to zoom out

This helps you verify focus on eyes, fine detail, or unwanted motion blur—before leaving the scene.

6. Audio Workaround: Get Great Sound Without a Mic Jack

Yes, the T7 has no microphone port. But that doesn't mean you're stuck with bad sound.

2 Audio Workarounds:

1. **Use a Smartphone + Lavalier Mic**
   - Clip a lav mic to your subject
   - Record using your phone (Voice Memos, Dolby On, or Recorder app)
   - Sync later in editing
2. **Use a Portable Recorder (Zoom H1n, Tascam DR-05)**

  - Standalone audio recorders capture crisp sound
  - Place close to the subject and sync later

7. Quick Menu Tips: Speed Up Your Workflow

Your **Q button** is a shortcut to everything that matters mid-shoot.

Use it to quickly:

- Change White Balance
- Switch Picture Styles
- Adjust AF Mode or Point
- Set Drive Mode (single, burst, timer)
- Toggle Metering Modes

Use this instead of digging through the full menu—especially during fast-paced shoots.

8. Grid Display: Composition Made Easy

Use the built-in grid to align horizons, apply the rule of thirds, or frame architecture.

How to Enable:

- **Menu > Shooting Settings > Grid Display → Enable**

This overlay appears in **Live View**—not the optical viewfinder.

9. Use the Rebel T7 for Streaming (Clean HDMI Trick)

Though not designed for it, the Rebel T7 *can* be used for **live streaming** with the right setup.

What You'll Need:

- Micro USB cable
- Canon EOS Webcam Utility (free from Canon's website)
- A tripod and good lighting
- Optional: External mic connected to your computer

Limitations: No clean HDMI out—but the EOS Webcam Utility bypasses that by feeding video directly via USB.

Great for Zoom calls, online teaching, or live YouTube sessions.

10. Self-Timer + Mirror Lock-Up = Sharper Shots

Shooting a long exposure or tripod scene? Avoid blur from pressing the shutter.

⊡ How:

1. **Enable Mirror Lock-Up** in Custom Functions
2. Use **2-Second Self-Timer Mode**

This delays the shutter after lifting the mirror, reducing vibrations that can cause softness in critical shots (like landscapes or macro work).

Final Thought: The More You Know, the More You Create

These aren't just party tricks.
They're the subtle enhancements that make your camera feel like an extension of your mind—not just a machine.

When you know where the hidden gems are, you stop working *against* your gear…
And start using it like it was custom-made for you.

Chapter 12

Growing Beyond Auto: Learn Like a Photographer

You've made it far. You're no longer just pressing buttons and hoping for the best.
You've learned how to see light, shape a scene, and command your Canon Rebel T7 with confidence.

But now comes the *real* turning point:

Shifting from *camera user* to *photographer*.

This chapter is your blueprint for growth.
It's where we:

- Develop your photographic eye
- Build practical exercises to strengthen your skills
- Learn how to analyze photos like a pro
- Discover when—and how—to upgrade your gear
- And most importantly, understand *why* you shoot, not just how

Because real photographers don't just *know* their gear...
They know themselves.

Step 1: Train Your Eye—Photography Is Seeing

Most people look at a scene and try to capture what they see.
Photographers look at a scene and decide *what to say.*

> **The camera doesn't take great photos. *You do—by noticing what others miss.***

Practice: "Frame Within the Frame"

Look through your viewfinder and:

- Focus only on a small detail—light on a face, a shadow, a line
- Ask: *What emotion or mood does this light convey?*
- Try shooting that same subject three different ways

□ ***Lesson: Composition isn't about adding more. It's about choosing what not to include.***

Step 2: Build Muscle Memory With Shooting Challenges

You learn fastest when you shoot with *constraints.*
Here's a set of **deliberate mini-challenges** that will supercharge your skills in just weeks.

◆ 7-Day Photography Challenge

Day	Focus	Goal
1	Light & Shadow	Shoot only in black & white
2	Texture & Detail	Get close—macro or tight frame
3	Motion	Capture blur *or* freeze motion
4	Silhouette	Shoot into the light
5	Portraits	Use window light only
6	Color Contrast	Capture two opposing colors
7	Storytelling	5 photos that tell a mini-story

Each day, review your shots. Ask:

- What works?
- What's missing?
- What could I do differently next time?

□These reflections matter more than any setting.

Step 3: Critique Your Work Like a Professional

Looking at your own images with intention is how you grow. Here's a checklist to guide your critique:

1. **What's the subject?**
   Is it immediately clear what the photo is about?
2. **Where does the eye go first?**
   Is it the brightest spot? The sharpest detail?
3. **Does the background support or distract?**

4. **Is the light working *with* or *against* the subject?**

5. **What emotion or atmosphere does it carry?**

⍈*Bonus Tip:* Show your work to non-photographers. Their gut reactions are often more useful than technical feedback.

Step 4: When (and What) to Upgrade

Your Rebel T7 can take world-class images.
But eventually, your skill will outgrow its limitations.
Here's when you *know* it's time to upgrade gear:

You Might Need a New Lens If:

- You want **blurry backgrounds** → Get a 50mm f/1.8 ("nifty fifty")
- You want **wide landscapes** → Look into 10–18mm EF-S lenses
- You want **sharper detail** → Upgrade from the kit lens to something faster or with better glass

You Might Need a New Camera If:

- You want faster **autofocus in video**
- You need an **external mic port**
- You shoot **professionally** and need dual card slots or higher dynamic range

- You're outgrowing the **sensor limitations** for editing large RAW files

But don't rush.

Mastery comes from your eye, not your gear.

Step 5: Learn from the World—Join the Photography Community

Photography is both solitary and communal. When you find the right community, you accelerate your growth.

Where to Learn More:

- **YouTube Channels:**
  - Peter McKinnon (storytelling & mood)
  - Jessica Kobeissi (portrait breakdowns)
  - The Art of Photography (composition deep dives)
- **Books:**
  - *Understanding Exposure* by Bryan Peterson
  - *On Photography* by Susan Sontag
  - *The Photographer's Eye* by Michael Freeman
- **Online Forums & Groups:**
  - Canon Photography Facebook Groups
  - Reddit: r/photography

- Flickr: Canon Rebel groups

Bonus Tip: Start a "photo journal"—shoot one intentional photo a day and caption it with a feeling or story.

Final Word: What You Create Is Worth Creating

You started this journey with curiosity.
Now you hold something much greater: **creative control.**

You've learned:

- To see with clarity
- To shoot with intention
- To use your camera not as a crutch but as a **companion**

Your photos don't have to be perfect.
They just have to be *real.*
Shot with attention, heart, and presence.

Don't chase perfection. Chase *presence.* Don't worry about the shot you missed. Worry about the story you can still tell.

Now go create something beautiful.

Chapter 13

Rebel T7 Photography Recipes & Inspiration

By now, your Canon EOS Rebel T7 isn't a stranger—it's a tool you've grown into.
But even with all the knowledge, it helps to have *go-to settings and creative ideas* you can rely on.

This chapter is your **quick-access photography cookbook.** Inside, you'll find:

- Ready-made **camera setting recipes** for specific scenarios
- Creative **shooting prompts** to inspire your next photo walk
- Challenges that keep your skills growing
- Confidence-boosting reminders to shoot even when you're uninspired

Whether you're packing for a trip, heading out with friends, or walking around your neighborhood, keep this chapter close—it'll spark ideas *and* get results.

Quick Camera Recipes for Everyday Scenes

Use these settings as a starting point. Adjust based on light and lens.

1. Blurry Background Portrait

For soft, dreamy background and focused subject

- **Mode:** Av
- **Aperture:** f/2.8 – f/4
- **ISO:** 100–400
- **Focus:** One-Shot + manual point on eye
- **White Balance:** Cloudy
- **Lens:** 50mm f/1.8 recommended

2. Ultra Sharp Landscape

For edge-to-edge clarity and dramatic depth

- **Mode:** Av
- **Aperture:** f/8 – f/11
- **ISO:** 100
- **White Balance:** Daylight
- **Use Tripod:** Yes
- **Focus:** Center third of scene

3. Moving Subject (Dance, Sports, Kids)

Freeze motion without blur

- **Mode:** Tv
- **Shutter Speed:** 1/500–1/1000
- **ISO:** 400–1600
- **Focus:** AI Servo + Continuous drive
- **White Balance:** Auto or Daylight

4. Night Street Scene Without Flash

For moody, low-light city photos

- **Mode:** M
- **Shutter:** 1/10–1 sec
- **Aperture:** f/4 – f/5.6
- **ISO:** 800–1600
- **White Balance:** Tungsten or Custom
- **Tripod:** Strongly recommended
- **Focus:** Manual focus or magnify with Live View

5. Golden Hour Glow

For warm, cinematic light at sunset

- **Mode:** Av
- **Aperture:** f/5.6
- **ISO:** 100–200
- **White Balance:** Cloudy or Shade
- **Exposure Comp:** -0.3 to deepen color
- **Picture Style:** Portrait or Neutral

6. Black & White Mood Shot

Use contrast and light shape for emotion

- **Mode:** M or Av
- **ISO:** 100
- **White Balance:** Auto
- **Picture Style:** Monochrome
- **Aperture:** f/4 – f/8
- **Focus:** Center point + recomposition

Personal Projects to Keep You Growing

⊡ 1. Photo Diary

Capture one image per day for a month. Print or create a digital zine.

□ 2. Video Memory Capsule

Film 3–5 second clips of life moments for one week. Edit into a 1-minute memory montage.

□ 3. Create a Themed Photo Box

Pick a theme (e.g., "quiet mornings" or "blue shadows"). Capture 10–20 images and curate a digital or printed collection.

Final Word: Keep Shooting, Keep Growing

Perfection isn't the goal.
Presence is.
Your Rebel T7 doesn't need to be the most expensive camera—it just needs to be *used with intention.*

Your eyes are learning to see.
Your hands are learning to respond.
And your stories are getting clearer, one frame at a time.

Chapter 14

Caring for Your Camera

Your Canon EOS Rebel T7 is more than a camera—it's your creative partner. But like any tool, it needs attention, care, and the occasional reset to keep performing at its best.

In this chapter, you'll learn:

- How to clean your camera safely (and what *not* to do)
- Best practices for storing and transporting your gear
- Signs of wear you should pay attention to
- How to extend your battery life and avoid SD card headaches
- Simple habits to keep your images—and your camera—sharp for years

Taking care of your gear doesn't have to be technical or tedious. In fact, it's one of the simplest ways to respect the creative process—and ensure your camera is ready whenever inspiration strikes.

1. Cleaning Your Camera (The Safe Way)

The Rebel T7 is a precision machine. Even small scratches or dust particles can affect image quality. Let's keep things spotless—but smartly.

⊡ What You'll Need:

- **Soft microfiber cloth** (NOT your shirt)
- **Lens blower** (hand-squeeze air blower)
- **Lens brush** (camel-hair or synthetic)
- **Lens cleaning fluid + tissue or cloth** (for glass surfaces only)
- **Sensor cleaning swab** (optional, for advanced users only)

⊡ Cleaning the Lens (Front & Rear Elements)

1. **Blow off dust** first—never wipe without removing grit
2. Gently brush the surface with a soft lens brush
3. If smudged, apply a drop of cleaning fluid to a microfiber cloth—not directly on the lens
4. Use gentle circular motion from center outward
5. Clean the rear glass only when removing the lens

Never use paper towels, tissues, or your breath. They leave residue or micro-scratches.

Cleaning the Camera Body

- Use a **dry microfiber cloth** for buttons and grip
- Remove dirt around buttons with a small brush or soft toothbrush
- Avoid moisture near the viewfinder, ports, or LCD

Tip: Gaffer tape can remove sticky residue or dust from crevices safely

Cleaning the Sensor (Only If Needed)

If you see dark spots in the same place on every image (especially at f/11 or higher), you may have sensor dust.

First:
Try the automatic sensor cleaner:

- **Menu > Sensor Cleaning > Clean Now**

Still there?
You may need a manual clean.

⊡ Sensor cleaning is delicate. If you're unsure, take it to a camera shop.
If you attempt it:

- Use **sensor swabs** and sensor-safe fluid
- Power off camera, remove lens, and clean gently with steady hand
- Only in a dust-free environment

2. Battery Health & Charging Habits

Your Rebel T7 uses the **LP-E10 battery**. It's reliable—but can degrade over time if not cared for.

⊡ Battery Tips:

- **Don't store batteries fully depleted**
- **Charge before long periods of non-use**
- **Avoid leaving the battery in the camera for weeks** when unused
- Always use **Canon-approved or high-quality third-party chargers**
- Store extra batteries in a **cool, dry place**, ideally in plastic cases

⊡ Battery draining fast? That's often due to

- Live View overuse
- High ISO and screen brightness
- Cold weather

Keep a **spare battery** with you—especially when traveling or doing long shoots.

3. Memory Card Care: Protect Your Photos

⊡ Best Practices:

- Use **Class 10 UHS-I SD cards** (32GB–128GB ideal)
- Always **format in-camera**, not on your computer
- Never remove card while camera is on
- Don't fill the card completely—leave 10–15% free space
- Use a **card reader**, not a USB cable, for faster, safer transfers

⊡ Signs Your Card Might Be Failing:

- Camera freezes or won't write
- Slow performance when reviewing images
- Image errors or corrupt files

If any of these happen, **stop using the card immediately** and back up your files.

4. Storing & Transporting Your Camera Safely

- Always **cap your lens** and turn camera off before storing
- Use a **padded camera bag**—never toss it into a backpack unprotected
- Avoid extreme temperatures, humidity, or sand (especially in beach environments)
- Store with a **silica gel packet** to absorb moisture
- Don't store near magnets or electronics that emit strong signals (they can corrupt SD cards)

Tip: For long-term storage, remove the battery and card, and keep the camera loosely wrapped in a soft cloth or pouch.

5. Regular Maintenance Schedule

Task	How Often
Wipe down camera body	After every use
Clean lens front element	Weekly or as needed
Format memory card	After backing up photos
Update firmware	Every 6 months (if available)
Deep clean sensor (if needed)	Every 6–12 months or when visible spots appear
Replace worn lens caps, hoods, or straps	As needed

Rule of Thumb: Treat your gear like an instrument, not an appliance. Respect it, and it will serve you well for years.

Final Word: When You Care for Your Gear, You Care for Your Art

Your Rebel T7 doesn't demand much—but it deserves attention. These small routines—wiping a lens, formatting a card, storing it correctly—might seem boring.

But together, they keep your camera ready for **what really matters:**

- The fleeting smile of a loved one
- A perfect sunrise
- A single, unrepeatable moment that tells your story

You've put time into learning this tool.
Now treat it like the creative ally it is.

Chapter 15

PROFESSIONAL SECRETS & QUICK FIXES

You've come a long way.
You understand your Canon EOS Rebel T7 inside and out—from settings to scenes, care to creativity.

Now, in this final chapter, we're going to tie everything together with

- **Real advice from professional photographers**
- **Fixes for common beginner mistakes**
- **Mindset shifts that elevate your shooting forever**

This is the chapter you'll return to whenever you feel stuck or unsure.
Not just to remember *how* to fix something, but to reconnect with *why* you're doing it in the first place.

⊡ 1. Think Like a Pro: 7 Simple Truths Most Beginners Never Hear

These aren't settings.
They're philosophies that shape how you approach every image.

⊡ 1. Get the shot. Then get the perfect shot.

Don't fiddle with settings and miss the moment.
Take the photo—then refine and reshoot if time allows.

⊡ 2. Light beats gear. Always.

Even the best camera won't save a poorly lit scene.
Look for soft, directional light. Use windows, shades, or even white walls.

⊡ 3. Composition is your superpower.

Where you stand and what you include or exclude matters more than your lens.
Use leading lines. Frame your subject. Simplify.

⊡ 4. Practice > Upgrades

A new lens won't fix bad technique.
Shoot every day. One hour with intention beats 100 shots without it.

⊡ 5. Shoot for feel, not likes.

Don't chase trends. Chase honesty.
A photo that *moves* you is always a good photo.

☐ 6. Know when to break the rules.

Rules like the rule of thirds, center framing, symmetry, or blur exist to be played with.
Once you understand them, get bold and bend them.

☐ 7. You're allowed to call yourself a photographer.

You don't need a portfolio or a client.
If you *see* and you shoot, you're in the club. Welcome.

2. Instant Fixes to Common Problems

"My photos are blurry."

Fix:

- Raise shutter speed (1/125 minimum)
- Enable Image Stabilization
- Use proper hand-holding posture
- Focus on the *eye* manually
- Use back-button focus for recomposition

"My images are too dark or too bright."

Fix:

- Watch the light meter (aim for 0 to +1)

- Use Exposure Compensation (+/-)
- Adjust ISO before touching aperture/shutter
- Use Manual Mode with Exposure Simulation

"My colors look wrong."

Fix:

- Set proper White Balance (avoid Auto in weird lighting)
- Try Cloudy for warmth outdoors
- Avoid harsh overhead lights indoors—use natural light

"My subject is in focus, but the photo still feels off."

Fix:

- Check the background—too busy? Too distracting?
- Shift angle or perspective
- Adjust composition—get closer, frame tighter

"Traveling soon—what should I pack?"

Pro Tip:

- Camera + 1 lens (18–55mm or 50mm)
- Extra battery
- Extra memory card

- Microfiber cloth
- Light bag or strap
- USB SD card reader

Leave the rest. Travel light. Focus more on moments, less on gear.

3. Build a Pre-Shoot Mental Checklist

Every professional walks through a quick process before they press the shutter.

Use this mental habit:

⊡ **What's my light source?**
⊡ **What's my subject?**
⊡ **Where do I want the eye to go first?**
⊡ **What's my message or emotion?**
⊡ **Did I check my focus point and exposure?**
⊡ **Do I want to capture this or interpret it?**

This alone will improve every image you take—regardless of your gear.

⊡ 4. Your Creative License

Photography is personal. Emotional. Messy. Surprising.

You will make mistakes.
You will shoot entire sessions that don't work.
You will doubt your eye, your edits, your gear, and yourself.

That's all part of the deal. And it's *exactly* what makes it beautiful.

Your Canon EOS Rebel T7 is a vehicle. But *you* are the engine.
Your voice. Your perspective. Your growth.
That's what matters.

So shoot when the light is perfect—and when it's not.
Shoot when you're inspired—and especially when you're not.
Shoot for clients. Shoot for strangers.
But always—**shoot for you.**

⊡ Final Words: You're Ready

You've completed this journey.
You've built a foundation that many photographers take *years* to find.

And from here, there are no limits.

> Your eye will evolve.
> Your courage will grow.
> And your work will speak for you—louder than any caption ever could.

So charge your battery.
Format your card.
Step outside.
And create something unforgettable.

You're not just using a camera now.
You're making something *only you* can make.

Keep going.

Acknowledgements

Creating this book has been a journey as much about photography as it has been about people—those who stood beside me, believed in the vision, and helped bring it into focus.

First, to the readers—whether you're picking up a camera for the first time or rediscovering your creative spark—thank you. Your curiosity, your hunger to learn, and your desire to capture the world around you are the very reasons this guide exists.

To the photographers who came before me—both legendary and unknown—your work inspired me to see light and shadow differently. You reminded me that the camera is not just a tool but a language.

To my early mentors and friends in the field, thank you for the countless late-night conversations, critique sessions, and shared sunrise shoots. Your insight, encouragement, and challenges helped me evolve from a hobbyist to a teacher.

A special thanks to the online photography communities, forums, and students I've taught—your questions, frustrations, and "aha" moments have shaped the clarity and heart of this guide. You reminded me what it feels like to start at square one.

To the editors, collaborators, and publishing team who helped bring this book to life with polish and purpose—your expertise is deeply appreciated.

Finally, to my family and loved ones—thank you for your patience when I disappeared behind the viewfinder or spent hours writing and refining pages. Your quiet support gave me the space to create freely and finish what I started.

This book is for everyone who ever doubted their creative potential. Here's proof that you don't need expensive gear or years of training—just a desire to learn, a willingness to try, and a spark of imagination.

Thank you for allowing me to be part of your photographic journey.

About the Author

Marc T. Billy is a passionate photographer, educator, and visual storyteller with over a decade of hands-on experience capturing the world through the lens. Known for his ability to break down complex photography concepts into simple, actionable steps, Marchas helped thousands of beginners transform their photography from blurry snapshots to frame-worthy masterpieces.

Marc's journey began not in a studio but in everyday life, where he taught himself the art of natural lighting, composition, and camera mastery through trial, error, and relentless curiosity. His work spans portraiture, landscapes, urban scenes, and lifestyle photography, often focused on finding beauty in ordinary moments.

Driven by a desire to make the Canon EOS Rebel T7 Guide accessible to everyone, Marc created the *Canon EOS Rebel T7 Guide* to empower new camera owners, hobbyists, and visual creatives to shoot with confidence—without the overwhelm of technical jargon.

When he's not behind the camera or mentoring aspiring photographers, Marc enjoys exploring hidden locations, experimenting with creative photo challenges, and turning everyday light into extraordinary images.

Marc believes that with the right guidance, anyone can master the camera and tell powerful stories through photographs. His mission is simple: to help you unlock your inner photographer—one frame at a time.

Made in the USA
Monee, IL
05 November 2026

67809037R00069